The TPM Playbook

A Step-by-Step Guideline for the Lean Practitioner

The LEAN Playbook Series

The 5S Playbook: A Step-by-Step Guideline for the Lean Practitioner
Chris A. Ortiz

The Kanban Playbook: A Step-by-Step Guideline for the Lean Practitioner
Chris A. Ortiz

The Quick Changeover Playbook: A Step-by-Step Guideline for the Lean Practitioner
Chris A. Ortiz

The TPM Playbook: A Step-by-Step Guideline for the Lean Practitioner
Chris A. Ortiz

The Cell Manufacturing Playbook: A Step-by-Step Guideline for the Lean Practitioner
Chris A. Ortiz

The LEAN Playbook Series

The TPM Playbook

A Step-by-Step Guideline for the Lean Practitioner

Chris A. Ortiz

CRC Press
Taylor & Francis Group
Boca Raton London New York

CRC Press is an imprint of the
Taylor & Francis Group, an **informa** business

A PRODUCTIVITY PRESS BOOK

CRC Press
Taylor & Francis Group
6000 Broken Sound Parkway NW, Suite 300
Boca Raton, FL 33487-2742

© 2016 by Taylor & Francis Group, LLC
CRC Press is an imprint of Taylor & Francis Group, an Informa business

No claim to original U.S. Government works

Printed on acid-free paper
Version Date: 20150806

International Standard Book Number-13: 978-1-4987-4166-8 (Paperback)

This book contains information obtained from authentic and highly regarded sources. Reasonable efforts have been made to publish reliable data and information, but the author and publisher cannot assume responsibility for the validity of all materials or the consequences of their use. The authors and publishers have attempted to trace the copyright holders of all material reproduced in this publication and apologize to copyright holders if permission to publish in this form has not been obtained. If any copyright material has not been acknowledged please write and let us know so we may rectify in any future reprint.

Except as permitted under U.S. Copyright Law, no part of this book may be reprinted, reproduced, transmitted, or utilized in any form by any electronic, mechanical, or other means, now known or hereafter invented, including photocopying, microfilming, and recording, or in any information storage or retrieval system, without written permission from the publishers.

For permission to photocopy or use material electronically from this work, please access www.copyright.com (http://www.copyright.com/) or contact the Copyright Clearance Center, Inc. (CCC), 222 Rosewood Drive, Danvers, MA 01923, 978-750-8400. CCC is a not-for-profit organization that provides licenses and registration for a variety of users. For organizations that have been granted a photocopy license by the CCC, a separate system of payment has been arranged.

Trademark Notice: Product or corporate names may be trademarks or registered trademarks, and are used only for identification and explanation without intent to infringe.

Library of Congress Cataloging-in-Publication Data

Ortiz, Chris A.
  The TPM playbook : a step-by-step guideline for the lean practitioner / author, Chris A. Ortiz.
    pages cm
  Includes index.
  ISBN 978-1-4987-4166-8 (acid-free paper) 1. Total productive maintenance--Handbooks, manuals, etc. I. Title.

TS174.O78 2016
658.2'02--dc23   2015026073

Visit the Taylor & Francis Web site at
http://www.taylorandfrancis.com

and the CRC Press Web site at
http://www.crcpress.com

Contents

How to Use This Playbook vii
Introduction ix

1 Start with 5S 1
    Introduction 1
    Red Tagging 2
    Why Sort? 3
    Painted Work Areas and Aisle Ways 6
    Machine Shop 7

2 Maintenance Inventory 13
    Introduction 13
    Benefits of Kanban 13
    Creating a Kanban Sizing Report 14
    Cycle Counting, Part Identification, Vendors, and On-Hand Inventory 15
    Identification of New Inventory Quantities 16
    Calculating New On-Hand Costs 17
    Creating a Kanban Sizing Report 17
    Break Down Items into Categories 18
    Maximum Quantity Examples 19
    Maintenance Kanban Cards 20

3 Baseline Your Equipment 23
    Introduction 23
    Seven Types of Abnormalities 23
    One-Turn Method 25

4 TPM Procedures 29
    Introduction 29
    Proactive Maintenance 30
    Preventive Maintenance 30
    Predictive Maintenance 30
    Creating Machine Operator Procedures 31
    Sample Icons for Procedure 31
    Solution-Based Engineering Procedure 36

5	**Visual TPM Boards** 39
6	**Conclusion** 43

Definition of Terms 45

Index 47

About the Author.. 51

How to Use This Playbook

In most cases, a playbook is a spiral-bound notebook that outlines a strategy for a sport or a game. Whether a football game, a video game, or even a board game, playbooks are all around us, and when written properly, they provide immediate and easily understood direction. Playbooks can also provide general information; then, it is up to the user of the playbook to tailor fit it to their individual needs.

Playbooks contain pictures, diagrams, quick reference, definitions, and often step-by-step illustrations to explain certain parts. Playbooks either can help you understand the entire game or can help you pick and choose to focus on one element. The bottom line is that any playbook should be easy to read and to the point and contain very little to no filler information.

The *TPM Playbook* is written for the Lean practitioner and facilitator. Like a football coach, a facilitator can use this playbook for quick reference and then be able to convey what is needed easily. If for some reason the person leading the actual TPM (Total Productive Maintenance) implementation forgets a "play," the person can reference the playbook. Either you can follow page by page and use it to facilitate a TPM implementation or you can go directly to certain topics and use it to help you implement that particular play.

Total Productive Maintenance

Total Productive Maintenance is a companywide approach to improving the effectiveness and longevity of equipment and machines. Depending on how automated your production processes are, TPM could play a major role in your Lean journey. TPM is a critical component of production line success to ensure product flow; equipment uptime is not disrupted and is operating at optimal levels.

Regardless of proactive measures, random breakdowns may occur. The goal here is to minimize their occurrence, with most of the maintenance conducted during scheduled downtime. Part of a comprehensive TPM program is creating accountability and ownership at all levels to help change the mindset of the machine operators and maintenance workers. Both are responsible for maintaining equipment, and each group plays its own role. We often see a laissez-faire approach from maintenance departments in traditional manufacturing companies, where machine maintenance is reactive. Without an organized and structured TPM program, maintenance staff can develop a feeling of invincibility in which they are the lifeline of the company. This is not good. They play a vital role in the overall health of the company, but they must work according to certain standards and guidelines just like those we place on production and machine operators.

This playbook bridges these common gaps.

Introduction

At first glance, the improvement techniques within the Lean philosophy appear to provide a solution to many types of production-related issues. A powerful and effective improvement philosophy, Lean can prevent company failure or launch an organization into world-class operational excellence.

I have been a Lean practitioner for over 15 years and have been involved in many Lean transformations. It does not matter the industry you work in, the product you produce, and even the processes your company uses to transform a finished good—the problems and opportunities you face are the same as those for everyone else. Your company is not "different" or the exception to every other company. You, as a Lean practitioner, desire a smoother-running facility, reduced lead times, more capacity, improved productivity, flexible processes, usable floor space, reduced inventory, and so on. Organizations implement Lean to make localized improvements or they can make Lean transform the entire culture of the business. Regardless of your aspirations and goals for Lean, you and many other companies face another similar situation: getting out of what I call *boardroom Lean* and moving toward implementation.

Have no illusions: Lean is about rolling your sleeves up, getting dirty, and making change. True change comes on the production floor, in the maintenance shop, in all the other areas of the organization and in implementing the concepts of Lean. Companies often become stuck in endless cycles of training and planning, with no implementation ever happening. This playbook is your guideline for implementation and is written for the pure Lean practitioner looking for a training tool and a guideline that can be used in the work area while improvements are being conducted. There is no book, manual, or reference guide that provides color images and detailed step-by-step guidelines on how to properly begin your TPM journey. The *Total Productive Maintenance Playbook* is not a traditional book, as you can probably see. It is not intended to be read like another Lean business book. The images in this playbook are from real TPM implementations, and I use a combination of short paragraphs and bulleted descriptions to walk you through how to effectively implement TPM.

Little or no time is wasted on high-level theory, although an introductory portion is dedicated to the 8 Wastes and Lean metrics. An understanding of wastes and metrics is needed to fully benefit from this playbook. This is not to imply that high-level theory or business strategies are not valuable; they are highly valuable. This playbook is for implementation so it will not contain filler information.

Chapter 1 covers the importance of implementing 5S (Sort, Set In Order, Scrub, Sanitize, and Sustain) and the visual workplace in your maintenance department. Without this foundation, TPM can be difficult to implement, and this chapter provides a detailed review of 5S but with a focus on how it is applied in a maintenance department.

Maintenance inventory control is a critical element of a Lean maintenance department. Spare parts, supplies, and general inventory related to maintenance functions can involve some of the highest costs in the department. To ensure higher productivity, it is important to organize and create a visual Kanban system for shop supplies and inventory. Using portions of the *Kanban Playbook*, Chapter 2 is a detailed review of Kanban and how it is applied in a maintenance department. Multiple examples, images, and illustrations are used for this purpose.

The next preparation step is to select a group of machines that will undergo the first TPM implementation and conduct what we call a baseline analysis on the status of the machines. Chapter 3 provides sample assessment forms that are used to identify what elements of the machine need to be brought up to "maintainable" level. This chapter then walks the reader through how to create action items and timelines to bring equipment up to a standard that will be a starting point for TPM. Machine modifications are discussed, as are safety improvements and changes that can be done to reduce maintenance technicians' time when conducting future TPM-related work.

Involving machine operators and maintenance together in your TPM program helps create an overall company approach to maintaining machines. Chapter 4 discusses how to establish their roles and responsibilities in the TPM program. Using pictures and real-life examples, the reader will be taught the best approach for selecting this work and how to start creating TPM procedures that will be used to illustrate and outline the work they will conduct. It will also cover the frequency of work, tools, material, and facility needs when machine operators perform their work. It is important to review current maintenance roles and decide if certain tasks can be passed on to machine operators or if other modifications are needed.

Visual TPM boards are the communication system that is used to convey critical information about the TPM work and the status of equipment. Chapter 5 walks you through how to design these boards using pictures and actual TPM boards in use.

8 Wastes

As a Lean practitioner and teacher myself, Lean manufacturing has been and will always be about waste reduction. Developing, sustaining, and improving on Lean will remove or reduce a significant amount of waste. Many of you reading this playbook already understand the concepts of waste and Lean. For those of you just getting started, here is a brief description of each waste:

Overproduction
Overprocessing
Waiting
Motion
Transportation
Inventory
Defects
Wasted human potential

Overproduction is the act of making more product than necessary, completing it faster than necessary, and making it before it is needed. Overproduced product takes up floor space, requires handling and storage, and could result in potential quality problems if the lot contains defects.

Overprocessing is the practice of extra steps, rechecking, reverifying, and outperforming work. Overprocessing is often conducted in fabrication departments when sanding, deburring, cleaning, or polishing is overperformed. Machines can also overprocess when they are not properly maintained and simply take more time to produce quality parts.

Waiting occurs when important information, tools, and supplies are not readily available, causing machines and people to be idle. Imbalances in workloads and cycle times between processes can also cause waiting.

Motion is the movement of people in and around the work area to look for tools, parts, information, people, and all necessary items that are not available. As a process contains a high level of motion, lead time increases and the focus on quality begins to decrease. All necessary items should be organized and be placed at the point of use so the worker can focus on the work at hand.

Transportation is the movement of parts and product throughout the facility. Often requiring a forklift, hand truck, or pallet jack, transportation exists when consuming processes are far away from each other and are not visible.

Inventory is a waste when manufacturers tie too much money into holding excessive levels of raw, work-in-process (WIP), and finished goods inventory.

Defects are any quality metric that causes rework, scrap, warranty claims, and rework hours from mistakes made in the factory.

Wasted human potential is the act of not properly utilizing employees to the best of their abilities. People are only as successful as the process they are given to work in. If a process inherently has motion, transportation, overprocessing, overproduction, periods of waiting, and defect creation, then that is exactly what the people will do—wasted human potential.

My hope is that you will read this playbook and not only be inspired but also be able to roll up your sleeves and begin your TPM journey after the last page is read.

Lean Metrics

To effectively measure your success with TPM and Lean in general, you need to establish a list of critical shop-floor metrics that can be measured and quantified. On the production floor, these metrics are often called key performance indicators (KPIs). Lean is a powerful improvement tool that can have a profound impact on reducing lead times, increasing output, improving productivity, and many other types of KPIs. In some cases, the change is dramatic. We recommend the following Lean metrics become part of measuring your overall Lean journey:

- Productivity
- Quality
- Inventory
- Floor space
- Travel distance
- Throughput time

Productivity

Productivity is measured in a variety of different ways. Productivity is improved when products are manufactured with less effort. This reduction in effort essentially is the reduction of waste. 5S is put in place to reduce or eliminate all of the steps and time associated with searching for items in an unorganized maintenance department. Also, a good TPM program will maximize the time of maintenance and machine operators. The 5S portion of TPM essentially clears the "smoke" of confusion in the work area and then provides a work environment that harnesses value-added work. Downtime is reduced on equipment while output increases when machines are working properly and not breaking down.

Quality

One of the goals of a TPM implementation is to improve and maintain the ability of the equipment to make high-quality parts on a consistent basis. By identifying the right machine improvements and having operators and maintenance personnel check critical attributes on a regular basis, defects, rework, and scrap can be reduced.

Also, through the implementation of 5S and Kanban in the maintenance department, tools last longer and work properly, and inventory is less likely to be damaged due to lack of organization.

Inventory

A lot of money is tied up in parts, material, and supplies that are used on a regular basis by maintenance staff. It has been my experience that many maintenance departments lack proper inventory control, run out of critical spare parts, waste time searching for parts, and the departments often become a dumping ground for the plants.

Maintenance tends to hold on to items for extended time periods for "just-in-case" scenarios. As 5S and Kanban are implemented, inventory costs go down, downtime on equipment decreases, and productivity increases in the department. Also, by reducing inventory levels, floor space opens up, allowing for further improvements to flow and travel distance in the department.

Inventory ties up money, contributes to clutter, takes up floor space, and often creates some of the most common physical obstacles in the company. Workers spend time shifting material and inventory around just to locate what they need. Time is lost by dealing with excessive inventory just to get to the items required to perform their work.

Floor Space

Floor space comes at a premium, and you need to start looking at the poor use of floor space as hurting the company's ability to grow. Maintenance departments need space to work on special projects and to conduct their TPM tasks. Floor space should not be used to store junk or to act as a collector of unneeded items. The implementation of 5S in maintenance can open up floor space and allow you to consolidate and rearrange the space better.

As a company becomes less organized and unneeded items begin to accumulate, more space becomes used for non-value-added items. This creates an increase in waste. Over time, items such as workbenches, garbage cans, chairs, unused equipment, tools, and tables tend to pile up, and valuable production space simply disappears. Rather than reduce waste and improve floor space use, the general approach is to add. Add building space, racks, and shelves, and you want to change your perception of space to better use, fewer non-value-added items, less waste, and less stuff.

5S is a powerful Lean tool that can improve the overall use of floor space in the maintenance department, and the examples in this playbook illustrate that.

Travel Distance

Here is the best way to view travel distance: The farther it has to go, the longer it is going to take. Long production processes can create a lot of waste and can reduce overall performance. Plus, longer-than-needed processes take up floor space. There are two ways to look at travel distance: the distance people walk and the distance inventory (product) is transported.

Travel distance is connected to overall lead times in a process and the entire factory. When WIP is created above required quantities, it takes up valuable floor space and increases the distance that the production line needs. As travel distance increases, floor space becomes improperly used, workers walk farther, and lead times are increased. Wait time in-between process also increases, and there is added lead time to maneuver inventory around.

When work areas are designed incorrectly, they can create a lot of walking for workers, and as the area becomes cluttered, more time is needed to find essential items to work.

Throughput Time

Sometimes used in conjunction with measuring travel distance reduction, throughput time is the time it takes the product to flow down the production process. Throughput time has a direct impact on delivery, and the longer it takes product to move through the plant, the longer it takes to be delivered. Throughput time in a maintenance department is the time it takes TPM tasks and other special projects to reach completion. A product is not completed down a production line, but the department is performing tasks that directly have an impact on the lead time in production. If equipment runs poorly or goes down, lead times are extended.

Improving these key Lean metrics and using them as a measurement of your success will have a profound impact on the overall financial success and long-term growth of the company. One could look at these Lean metrics simply as process metrics because they can be measured at the shop floor level. Production workers need to work in an efficient environment to be successful contributors to optimal cost, quality, and delivery. Each Lean metric improved complements another, and another, and so on. As you become better as a Lean practitioner, your understanding of how they relate to each other will become second nature.

Chapter 1
Start with 5S

Introduction

The successful implementation of total productive maintenance (TPM) into your organization will help change the way the culture thinks and perceives the company's important assets. Within Lean, there are tools and concepts for reducing waste. 5S is a great starting point, and the purpose of this chapter is to show you how to implement 5S in the maintenance department. If maintenance staff and operators are required to perform TPM activities, what is the state of organization in their work areas? 5S creates the foundation for continuous improvement, and TPM helps develop a sense of pride and discipline for the culture. You also want your maintenance staff to spend their time performing TPM and special projects, not searching for tools and parts in a cluttered work area.

Evaluate what is needed to perform the tasks in a work area and remove anything unnecessary. It is good practice to use an identification system while conducting the sorting step. This system is called *Red Tagging*.

- Parts
- Tools
- Workbenches
- Garbage cans
- Fixtures and jigs
- Documentation
- Supplies
- Equipment
- Anything that can be removed

Break down the sorted items into three categories:

1. **Garbage and junk (throw away or recycle).** There is not much point in Red Tagging garbage, but it still needs to be removed.
2. **Unneeded, never to return.** This category will have most of the Red Tags. Further in this book, I provide some options of how to deal with these items.
3. **Low-use Items.** Low-use items are essential but not used that often, maybe once a month or in wider time frames. Red Tag these items but place them in a separate pile. Just make sure when you are setting things in order they are organized away from daily use items.

Make sure to create an inventory list of all items to be removed to help in the final disposition and removal from the company.

Red Tagging

- Organized approach to sorting
- Keeps track of what is being removed
- Allows for efficient removal from facility
- Three parts to a Red Tag event:
    - Attaching Red Tags: Red Tag
    - Red Tag area (temporary staging): Red Tag area
    - Removal procedure

This is a Red Tag area.

Why Sort?

- Unneeded inventory takes up space.
- Extra parts require wasted transportation and motion.
- Without sorting, it is tougher to find what is needed.
- Unneeded equipment becomes obstacles.
- It is tough to bring in new product lines.
- It is tough to bring in new equipment.
- The confusion that clutter creates is cleared.
- Clarity is obtained on what is truly needed.
- It prepares for setting things in order. (Why organize items you do not use?)

Is everything here really needed?

Red Tag area: service shop sorted items

Maintenance parts room: sorted items

Maintenance Red Tag area: rented container

The second stage of 5S is *Set in Order*. Set in Order is the act of creating locations for all essential items needed in the work area. It is the act of organizing what is needed so it is easily identifiable in a designated place. During this phase, it is recommended that the implementation team work from the floor up, focusing on the layout of the maintenance department.

It is at this point in the implementation when work areas, aisleways, and floor locations are established. Once this is complete, then smaller items such as tools, supplies, fixtures, and parts are given home locations.

Everything has a home.

Painted Work Areas and Aisle Ways

Painted aisles in maintenance.

Roll paint works best, especially in dirtier work environments like a maintenance department where there is heavy traffic. Spray paint also works well, just make sure to purchase an industrial-strength paint.

Tape off line area and paint lines 2 to 4 inches thick.

Machine Shop

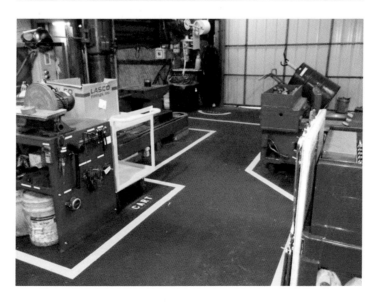

8 ■ *The TPM Playbook*

Maintenance equipment must have a home location.

Larger maintenance tools

Start with 5S ■ 9

Maintenance shadow board

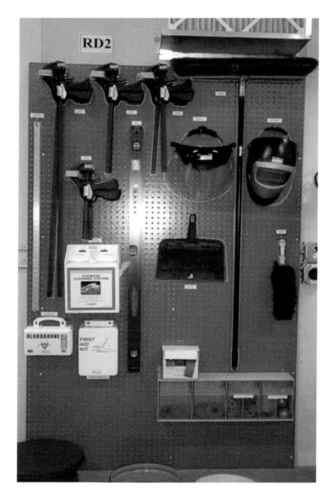

Maintenance tools and supplies

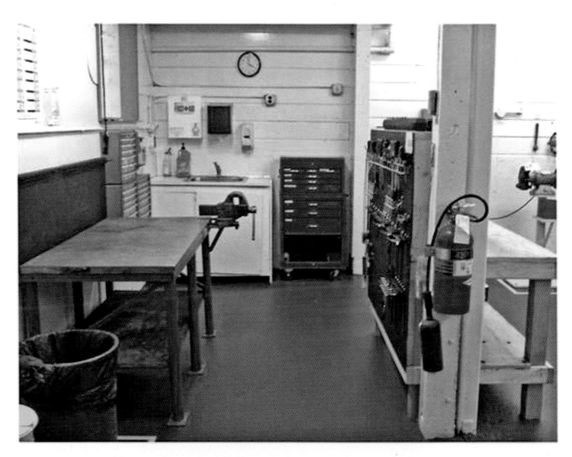

Maintenance shop at a cold storage facility

The third stage of a 5S implementation can be done after the Sorting or during the Set in Order phases. Scrubbing is the cleaning portion of the implementation. It is not intended to be an intensive clean, but here are some areas to focus while cleaning the area:

- Clean and degrease equipment
- Refill fluids as needed
- Wipe down work surfaces
- Wipe out storage bins
- Wipe shelves and racks
- Clean tools
- Wipe down garbage cans
- Sweep and mop the floor
- Painting equipment

Cleaning the floor

In many cases, it is good practice to paint equipment, tables, dollies, shelves, and even the outside of garbage cans during the Scrubbing portion of the project. It creates a showroom appearance and will also help make labels, stencils, and other designations stand out.

Painting hand trucks.

Machine shop before

After cleaning and painting

The concept of 5S Standardization is similar to how roadways and highways and all their visual markings are implemented and used in our everyday lives. As an example, the design of a stop sign is standard across all roadways in the United States.

The design and meaning of all visual roadway systems is identical to reduce confusion. It is recommended that your company come up with a 5S Standardization guideline for your implementation teams.

- Creates consistency in your 5S implementation
- Tool board consistency:
  - All tool boards are painted the same color.
  - Boards have shadows and tool labels.
  - Boards are identified with an address (M1, R3, etc.).

- Identify floor tape or paint colors to define categories:
  - Garbage cans **(green)**
  - Fixed items **(yellow)**
  - Finished goods **(black)**
  - Part locations **(blue)**

Once your implementation of the first four of the 5Ss is complete, you need to create a sustaining program that makes sense for your company and your culture. Every company is different, as is how each one establishes the guidelines and practices needed to sustain the improvements. Some companies can rely on the culture and have no real management systems in place. Others need formalized systems. Sustaining the 5S program is the hardest. Your sustaining efforts will never end, including continually improving on what was already implemented.

Each company must find its way with sustaining. Here are a few recommendations:

- Create an end-of-day cleanup procedure.
- Conduct a daily/shift walkthrough.
- Establish a 5S audit sheet.
- Create and maintain a 5S tracking sheet.

Chapter 2
Maintenance Inventory

Introduction

Kanban cards are a visual signal system that is implemented to trigger the need for inventory. These cards are placed in the maintenance area where inventory is stored and can have a significant impact on productivity. Cards can be placed into work areas and used when material and supplies are running low, and the card is turned in to a material handler for retrieving the needed item. The card contains all the vital information needed, including part number, part description, quantity, and locations.

Kanban card systems can also be used when ordering from outside suppliers, but this generally takes longer to implement. This chapter describes the internal Kanban card system within a company.

Benefits of Kanban

- Reduces inventory
- Reduces cost
- Reduces floor space use
- Reduces motion
- Creates better visibility of shortages
- Helps transform to a pull system
- Controls inventory
- Creates a robust and repeatable system

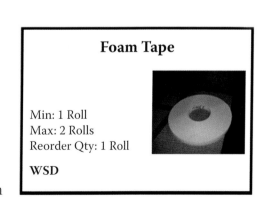

Foam Tape

Min: 1 Roll
Max: 2 Rolls
Reorder Qty: 1 Roll

WSD

Before you can run out and start implementing a Kanban system, there is some preliminary work that needs to be done. This chapter covers a variety of different planning items that will help implement a more effective and user-friendly Kanban system.

The following items are discussed:

- Creation of a Kanban sizing report
- Cycle counting, part identification, vendors, and on-hand inventory
- Identification of inventory quantities
- Calculating on-hand cost

Creating a Kanban Sizing Report

Maintenance Shop Supplies Report

Category	Part Description	Part #	Vendor	Cost per Unit	On Hand	Total on Hand
Fittings	#6 Quick Connects	FF-371-6FP	ACH	$7.50	13	$97.50
Fittings	#6 Quick Connects	FF-372-6FP	ACH	$7.50	13	$97.50
Fittings	Quick Disconnects	STUCCI-M-A7	ACH	$7.50	5	$37.50
Fittings	Quick Disconnects	STUCCI-F-A7	ACH	$7.50	4	$30.00
Fittings	Quick Disconnects	STUCCI-M-FIR614	ACH	$7.50	2	$15.00

A Kanban sizing report is a document that can be created in Microsoft Excel. It is used to perform cycle counting and to obtain baseline information for the implementation.

Category: Break down your inventory into categories based on the department

Part Description: Description of the supply part that is identified by the vendor

Part #: Actual part number used to order from vendor

Vendor: Name of supplier you order the part from

Cost per Unit: The average cost of the part from the supplier

On Hand: How many parts are on hand at the moment of cycle counting

Total Cost on Hand: Multiply the cost per unit times the on-hand quantity

Cycle Counting, Part Identification, Vendors, and On-Hand Inventory

Staff using Kanban sizing report to conduct cycle counts on maintenance supplies and parts

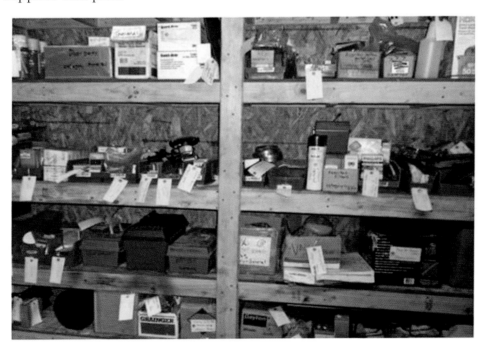

Tagging parts in the storage area with information on the part helps match the Kanban card with items during implementation.

Identification of New Inventory Quantities

Maintenance Shop Supplies Report

Category	Part Description	Part #	Vendor	Cost per Unit	On Hand	Total on Hand	Max Qty	Min Qty	Reorder Qty
Fittings	#6 Quick Connects	FF-371-6FP	ACH	$7.50	13	$97.50	4	1	3
Fittings	#6 Quick Connects	FF-372-6FP	ACH	$7.50	13	$97.50	4	1	3
Fittings	Quick Disconnects	STUCCI-M-A7	ACH	$7.50	5	$37.50	4	1	3
Fittings	Quick Disconnects	STUCCI-F-A7	ACH	$7.50	4	$30.00	3	1	2
Fittings	Quick Disconnects	STUCCI-M-FIR614	ACH	$7.50	2	$15.00	1	0	1

Some people think you need to look at a lot of data to establish new inventory levels. Depending on the situation, this might be necessary when it comes to production lines. This is an example from a maintenance department with sporadic usage information. Simply use the experienced people in the department to establish new inventory levels.

You will have to go through each part and discuss its usage, vendor, and lead time to come up with new maximum and minimum quantities. As odd as it may appear, this process is effective in the first pass, and you can refine your numbers after the implementation.

Max Qty: This establishes the highest level of inventory you want based on usage.

Min Qty: This is the quantity of safety stock to work from while the part is being ordered.

Reorder Qty: Once the minimum quantity is reached, this will trigger the need to turn in the Kanban card; this is the quantity that is ordered to refill to the required maximum.

Calculating New On-Hand Costs

Part #	Vendor	Cost per Unit	On Hand	Total on Hand	Max Qty	Min Qty	Reorder Qty	New Max Cost	Difference
FF-371-6FP	ACH	$7.50	13	$97.50	4	1	3	$30.00	$67.50
FF-372-6FP	ACH	$7.50	13	$97.50	4	1	3	$30.00	$67.50
STUCCI-M-A7	ACH	$7.50	5	$37.50	4	1	3	$30.00	$7.50
STUCCI-F-A7	ACH	$7.50	4	$30.00	3	1	2	$22.50	$7.50
STUCCI-M-FIR614	ACH	$7.50	2	$15.00	1	0	1	$7.50	$7.50
		Total On Hand Before		$277.50	Total On Hand Savings			$120.00	$157.50

This final Kanban sizing report shows the previous on-hand costs of inventory at $277.50 and the new on-hand cost after the implementation will be $120.00. Although this is a simple example, if you follow this thought pattern when designing your Kanban system, it could reach a savings of hundreds of thousands to millions of dollars depending on the level of inventory and cost per unit.

Creating a Kanban Sizing Report

There are much more complex Kanban sizing reports, and the number of parts will dictate how big the report will be. Now, this completed report is used to help you organize the storage and work area. Use the newly established maximum quantities to dictate the size of bins, shelf and work space, and overall floor space. Once that is complete, you can use the report to create a Kanban system.

Often, the information that is collected during the cycle counting can be obtained from some form of material information system, such as MRP (material requirement planning), ERP (enterprise resource planning), or some type of maintenance tracking software. If you have faith in the system's integrity, then go ahead. However, nothing provides any deeper insight on the current state, buying habits, size of parts, and unused inventory than sheet cycle counting.

Once you have completed your cycle counting, baseline, and the Kanban sizing report, you are ready to design the Kanban system for implementation.

Break Down Items into Categories

- Parts
  - Brackets
  - Panels
- Material
  - Sheets of metal
- Shop supplies
  - Rags, fillers, adhesives, cleaners, lubricants, tape
- Hardware
  - Nuts/bolts/washers, fasteners, and the like

Maintenance Inventory ■ 19

Maximum Quantity Examples

Parts room Kanban card

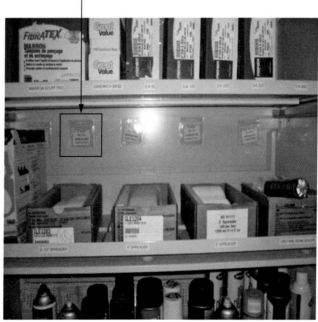

Maintenance Kanban Cards

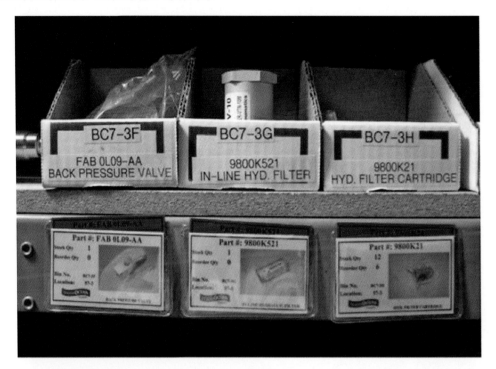

Cards contain pictures for quick reference.

Parts are organized based on machine type.

Companies often overlook the impact of poor inventory control and buying habits in a maintenance department. From a Lean perspective, a lot of time and frustration are associated with maintenance inventory, and gaining control of this part of the department will create a much more productive work area. There also are cost savings opportunities as well that can be captured; often, inventory budgets can be lowered, and that money can then be used for more important aspects of the department. Reinvestment into more maintenance staff, better equipment, and cross training the staff seems much more appropriate than buying excess supplies and spare parts.

Chapter 3

Baseline Your Equipment

Introduction

Equipment baselines are a great starting point for your total productive maintenance (TPM) program; the intent is for you to identify where current conditions are not being met on the machines. Before you can implement and execute ongoing preventive maintenance activities, you must identify certain repairs, upgrades, and modifications that are needed to bring the machines up to a "maintainable" level.

Seven Types of Abnormalities

- Minor flaws, parts, and vital components
- Basic conditions not fulfilled
- Inaccessible places
- Sources of contamination
- Sources of quality defect
- Unnecessary items
- Unsafe places or conditions

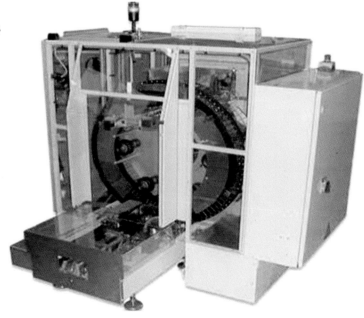

Machine Baseline Analysis Form

			Priority 1,2,3	Schedule Date	Percent Complete	Comments
Minor Flaws	A	Missing Paint/Rust				
1	B	Missing Decals				
	C	Scratches/Dings				
	D					
Parts and Vital Components	A	Cracking	Priority 1,2,3	Schedule Date	Percent Complete	Comments
2	B	Excessive Noise				
	C	Needs Repair or Replacement				
	D	Excessive Wear				
Basic Conditions not Fulfilled	A	Lubrication	Priority 1,2,3	Schedule Date	Percent Complete	Comments
3	B	Leveling/Upright/Rigidity				
	C	Loose Electrical Connections				
	D	Alignment/Adjustment				
Inaccessible Places	A	Special Tools Required	Priority 1,2,3	Schedule Date	Percent Complete	Comments
4	B	Poor Visibility				
	C	Awkward Positioning				
	D	Heavy Lifting				
Sources of Contamination	A	Seals/Gaskets	Priority 1,2,3	Schedule Date	Percent Complete	Comments
5	B	Fluid Replacement				
	C	Cleanliness				
	D	Filters				
Sources of Quality Defect	A	Needs Calibration	Priority 1,2,3	Schedule Date	Percent Complete	Comments
6	B	Machine Upgrades				
	C	Improper Materials				
	D	Insufficient Tooling				
Unnecessary Items	A	Tools	Priority 1,2,3	Schedule Date	Percent Complete	Comments
7	B	Personal Effects				
	C	Obsolete Information				
	D	Spare Parts				
Unsafe Places or Conditions	A	Ineffective Panels/Shrouds	Priority 1,2,3	Schedule Date	Percent Complete	Comments
8	B	Insufficient Signage				
	C	Maintenance Safeguards				
	D	Pinch-Points				

One-Turn Method

Another analysis you can perform as a complement to your baseline exercise is to conduct a one-turn method analysis. This analysis is something my team has done for years and has proven to be useful in reducing preventative maintenance (PM) cycle times. A one-turn method analysis is used to identify opportunities where typical turning operations can be reduced or eliminated. To simplify, examine how much time machine operators and maintenance staff spend in loosening and tightening hardware to remove panels and other items on the machine. Often, the time to remove and replace panels and enclosures is well more than the time to perform the PM tasks. Converting nuts, bolts, and screws to something that turns "once" can greatly reduce cycle time. Some form of quick disconnect system is ideal.

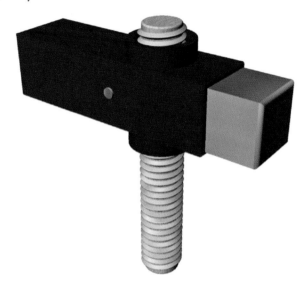

FAST–NUT	
Description: -The Fast-Nut needs no tools to install or remove from a threaded stud. -Install by first depressing the thread-engaugement-button. Then, slide the Fast-Nut over threaded stud, release the button, and tighten as required. -Remove by relieving the applied torque, depress the button and slide off.	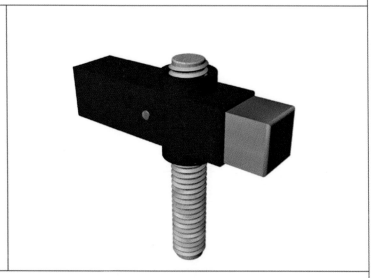
Purpose: -The Fast-Nut was created as a replacement for current nut-&-washer type fasteners. Very useful on machine access panels or anywhere a TPM activity requires removal of a cover. ***Metric Improvements:*** -No tools required for use. Highly decreased Cycle Time regardless of length of threaded stud on which the Fast-Nut is installed.	

Using manufacturing engineers and maintenance staff, come up with creative ways to reduce or eliminate these turning operations. You will find that your need for tools is also reduced or eliminated.

- Dovetails
- Clamps
- Sliding channels for panels
- Quick disconnect

You can add these one-turn method modifications to your list of tasks for baselining as well.

One-Turn Method Assessment

Machine/Equipment	Department/Area	Location on Machine	Purpose of Application	Recommended Change

Chapter 4

TPM Procedures

Introduction

Machine operators are the first line of defense and have firsthand knowledge of the equipment. Assigning them simple-to-moderate tasks in the total productive maintenance (TPM) journey is important. They can detect early signs of problems and perform simple work to ensure the machines they work on every day are working properly.

When creating procedures for machine operators, make sure the work is simple, not time consuming, and can be performed with ease. Make sure to make the procedure easy to read with lots of pictures, images, and symbols so almost anyone can understand the information.

Machine	DCM 5–10	
		Page 1 of 2
Sequence	**Work Content**	
1	Check Motor for Leaks	
2	Vacuum Debris and Dust	
3	Oil Rails	
4	Clean Windows Inside and Out	
Tool	**Qty**	**Facility Requirements**
Shop Vac	1	Electrical
Rags	1	EYE PROTECTION
DTE Lite Oil	1	
Shop Vac Extension	1	
Windex	1	

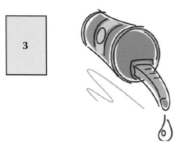 

To help you identify what machine operators and maintenance staff will be assigned to do, use the following three categories of TPM as a guideline:

- Proactive maintenance
- Preventive maintenance
- Predictive maintenance

Proactive Maintenance

- Checking gauges
- Cleaning equipment
- Filling fluids
- Tightening and loosening
- Simple and frequent

Preventive Maintenance

- Replacing parts
- Modifying equipment
- Tasks requiring special tools
- Tasks requiring shutdown
- Moderate and less frequent tasks

Predictive Maintenance

- Frequency is based on time or another method.
- Filters, motors, and belts are changed according to a certain time rotation.
- Changes are based on historical data.

Creating Machine Operator Procedures

- Checks should be at the simple-to-moderate level.
- Frequency is daily or monthly.
- Allow time in the day and week to perform the work.
- Provide all necessary tools and material in the work area.
- Organize tools and material with the 5S methodology.
- Use pictures and icons to create the procedure.
- Post the procedure and calendar of frequency on or near the equipment.

Sample Icons for Procedure

Clean windows or doors   Grease   Tape Measure

Vacuum   Tighten or adjust   Fill fluid

32 ◼ *The TPM Playbook*

The TPM implementation team reviewing machine operator roles.

This is an image used in creating TPM procedures for machine operators.

This image was also used in creating TPM procedures for machine operators.

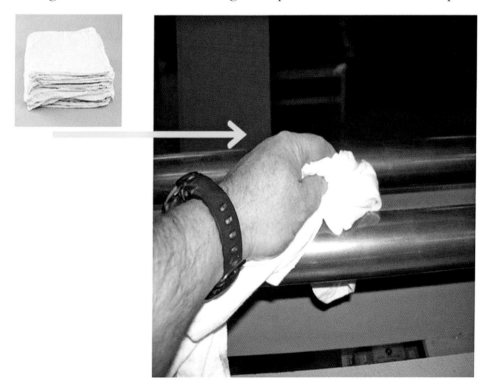

Here is a nice visual for wiping down rails.

34 ■ *The TPM Playbook*

This is an example of a lock-out tag out image in a TPM procedure.

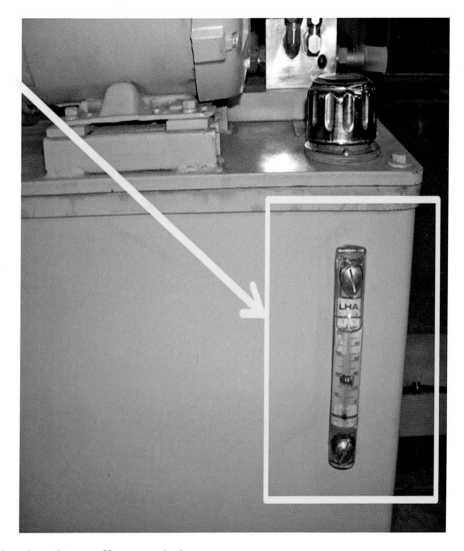

Check fluid level and top off as needed.

Solution-Based Engineering Procedure

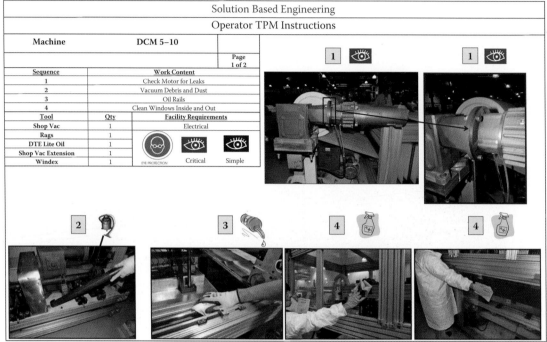

Maintenance staff are responsible for moderate-to-complex TPM tasks as their work may require shutting down the equipment, replacing parts, and modifications. The work they perform is less frequent, sometimes monthly or quarterly. Generally, their procedures are in the maintenance department, and they retrieve the procedure when they are grabbing tools and supplies for the work.

It is good practice during your TPM implementations to review current maintenance TPM activities and decide if some of the tasks can be passed on to machine operators. Make sure when you pass this work to a machine operator that the work is simple and quick. You want to place responsibility on the machine operators; just remember that their job is to run the equipment and make product first. Try to find a nice balance between departments.

TPM implementation team creating procedures.

TPM implementation team reviewing maintenance records.

38 ■ The TPM Playbook

The image that follows is a great example of a maintenance TPM procedure for a lumber mill. The saw change is performed less frequently and must be done when the machine is off. The proper shutdown procedure for changing the saw blades is outlined. This document is stored in the maintenance department and is used for reference when performing the work and for training new maintenance staff.

Production Saw Change SWP

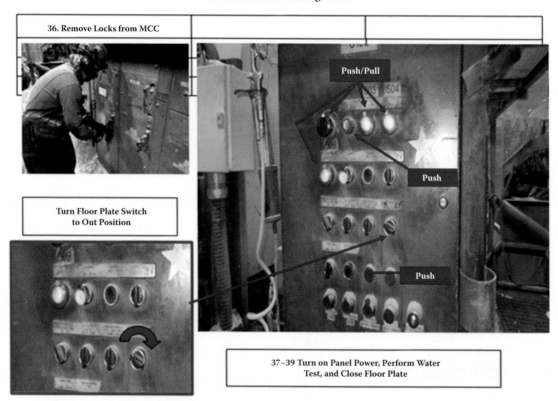

Chapter 5
Visual TPM Boards

Total productive maintenance (TPM) boards are a great mechanism for communicating vital information on the status of equipment in work areas. Their primary purpose is to allow machine operators and maintenance staff to visually send information back and forth to and from each other. The secondary purpose is for other departments and managers to have updates as they walk through the area. These boards are intended to provide quick and accurate information to all.

The TPM boards should be customized for the needs of the process, but there are certain aspects of each board that are needed to ensure it is used properly.

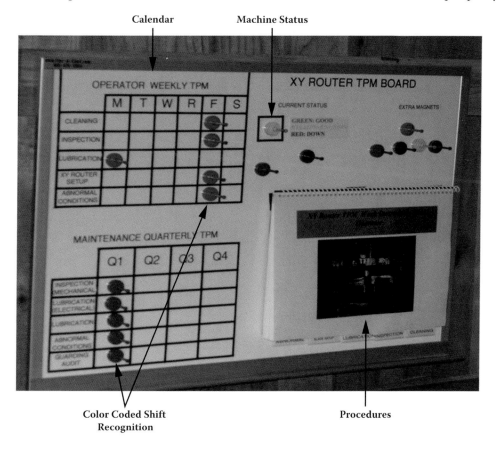

TPM boards need to have certain information to communicate effectively. Here are the key attributes of board design:

- Calendar of frequency for machine operators
- Calendar of frequency for maintenance staff
- Color magnets for shift recognition
- Machine status
- Section to place comments
- Machine operator TPM procedure
- Communication light (optional)

Create boards using the following:

- Use a magnetic dry erase board
- Use black pinstriping for borders
- Use a label maker for titles
- Mount the TPM board on or near the machine.

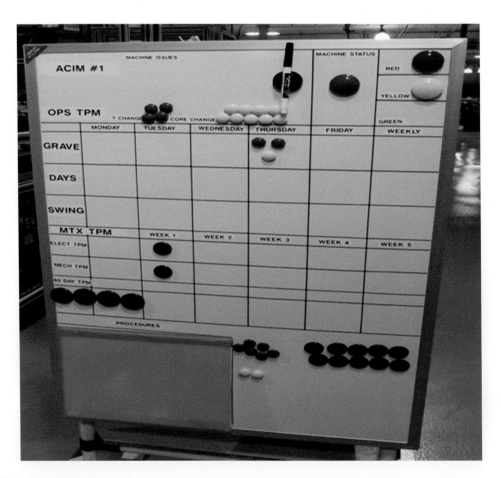

This is a TPM board in an Arizona company.

Visual TPM Boards ■ 41

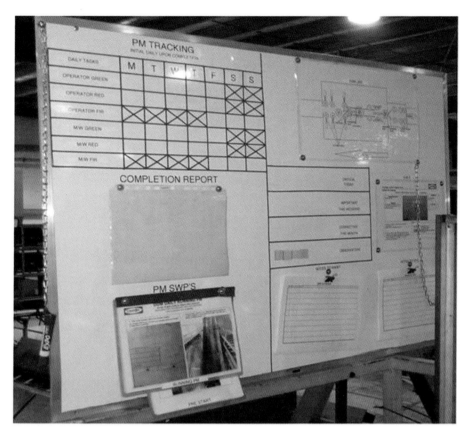

This TPM board is at a lumber mill in Canada.

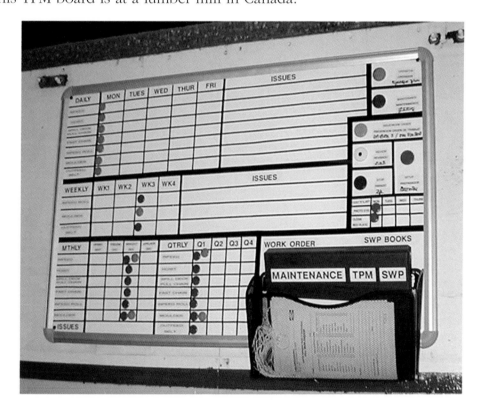

This TPM board is at a lumber mill in Washington.

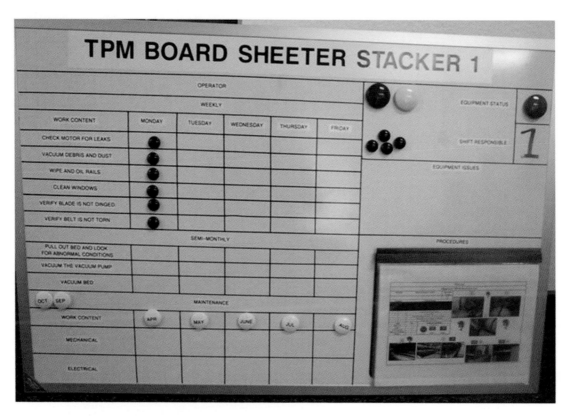

This TPM board is in a composites company.

Chapter 6
Conclusion

A companywide total productive maintenance (TPM) program can have a profound impact on your plant's performance. Bridging the potential gap between production and maintenance and creating a culture of teamwork and alignment will shift the mind-set for the better. It will reduce the amount of finger-pointing, and as a leader, it will show your commitment to the long-term health of your company's assets. Good luck!

Chris Ortiz

Definition of Terms

Daily walkthrough: Performed after the end-of-day cleanup, the walkthrough is conducted by a supervisor or worker to verify the cleanup is complete.

Defects: Mistakes made in the process requiring rework, material scrap, and lost products.

End-of-day cleanup procedure: A sustaining document that outlines the cleanup and reset requirements for the work area after each shift or day.

5S audit form: A scoring system used to rate the level of sustaining and used as a guideline for continuous improvement.

5S tracking sheet: A visual document posted in high-traffic areas that displays the scores from the 5S audit form.

5S and the visual workplace: Lean implementation concept of creating a highly organized work environment where everything has a place. Labels, designations, paint, and signage are used as examples to create the visual workplace.

Floor space: Performance measurement of how much factory space is being used to conduct value-added work. It is often measured in profit per square foot or revenue per square foot.

Inventory: Higher-than-needed inventory levels due to excessive purchasing of raw material, overproducing work in progress (WIP), and unsold finished goods. Inventory ties up working capital, takes up floor space, and adds to longer lead times.

Motion: Movement of workers generally leaving their work areas to find items unavailable there.

One-turn method: A concept of converting typical turning operations into quick disconnects to remove and replace equipment panels.

Overprocessing: The act of overperforming work steps such as redundant effort and extra steps.

Overproduction: The act of producing more product than necessary, performing work in the wrong order, and creating unneeded inventory.

Productivity: One of the six Lean metrics that is a measurement of a worker's efficiency in a process. Often, it is a comparison of the time allocated to perform work to the actual time the worker took to perform it.

Quality: Internal measurement of rework, scrap, and defects in a production process.

Red Tagging: An organized approach to sorting in which Red Tags are placed on items to designate them as unneeded. Red Tag items are placed in a staging area for permanent removal from the company.

Right sizing: Concept of customizing the work area to identify the minimum amount of space needed to store items.

Scrub: Act of cleaning and painting the work area to create a showroom condition.

Set in order: Act of complete organization of the company by which all items are given home locations.

Shadow board: A visual mechanism for organizing tools. Shadow boards provide instant feedback on home locations and missing tools and opens up floor space by eliminating the need for tool boxes and shelves.

Sort: Act of discarding and removing all unnecessary items from the work area.

Standardize: Act of creating consistency in the 5S implementation through guidelines for the visual workplace.

Sustain: The act of maintaining the work area after a 5S implementation.

Throughput time: Time associated with all value-added and non-value-added time in a process. It is the time it takes material to get through the first and last steps of the entire factory, from raw material to finished goods.

Total productive maintenance: A companywide approach to the effectiveness and longevity of equipment and machines.

Transportation: The movement of raw, WIP, and finished goods throughout the company.

Travel distance: Measurement of the physical distance product and workers go and the time associated with it. A long travel distance equates to longer lead times in the process.

Visual TPM boards: Communication boards commonly made from a dry erase board, pinstriping, and magnets; used to convey important information on the status of equipment and its maintenance tasks.

Waiting: When work comes to a stop due to lack of necessary tools, people, material, information, and parts. Wait time is often called queue time.

Wasted potential: Poor use of people, including skill sets not utilized, wrong job placement, and workers consumed in wasteful steps.

Index

A

abnormalities, 23
addresses, *see* Home locations
aisle ways, 6
analysis form, 24
Arizona, 40
assessment form, 27
audit form, 12, 45

B

baselining equipment
    abnormalities, 23
    analysis form, 24
    fast-nut example, 26
    one-turn method, 25–27
    overview, 23
    quick disconnect system, 25, 26
boards
    attributes, 40
    defined, 46
    examples, 40–42
    materials to create, 40
    overview, 39–42, 46
buying habits, 21

C

calendar, 39, 40
categories
    Kanban sizing report, 14
    maintenance inventory, 18
    sorting items, 2
    TPM guidelines, 30
clamps, 26
cleaning activities, *see* Scrubbing activities
cold storage facility, 10

color-coding
    standardization activities, 12
    visual TPM boards, 39, 40
comments, TPM boards, 40
communication lights, 40
component abnormality, 23
composites company, 42
conditions not filled, 23
consistency, 12
contamination sources, 23
costs
    calculation, on-hand inventory, 17
    Kanban sizing report, 14
    savings opportunities, 21
cross training, 21
culture, 1
current conditions
    baseline equipment, 23
    maintenance activities, 36
cycle counting, 15, 17

D

daily walkthrough, 12, 45
defects
    abnormality, 23
    defined, 45
    overview, *xi*
designations, 11
disconnect system, 25, 26
documentation, 2
dollies, 11
dovetails, 26

E

end-of-day cleanup procedure, 12, 45
equipment
    home locations, 8

investment in better, 21
preventive maintenance, 30
proactive maintenance, 30
Red Tagging, 2
scrubbing activities, 10–11
equipment, baseline
abnormality types, 23
analysis form, 24
assessment form, 27
one-turn method, 25–27
overview, 23

F

fast-nut example, 26
fingerpointing, 43
finished goods, 12
5S activities
audit form, 12, 45
floor space, *xiii*
machine shop, 7–12
overview, 1–2
painted work areas and aisle ways, 6
productivity, *xii*
quality, *xii*
Red Tagging, 2–3
sort activities, 3–5
sustain activities, 12, 46
tracking sheet, 45
visual workplace, 45
fixed items, 12
fixtures and jigs, 2, 5
flaws, 23
floor space
defined, 45
5S activities, *xiii*
locations establishment, 5
overview, *xiii*
scrubbing activities, 10–11
fluid levels
icon, 31
proactive maintenance, 30
scrubbing activities, 10
total productive maintenance procedures, 35
foam tape example, 13
frequency, predictive maintenance, 30

G

garbage and junk, 2
garbage cans, 10, 11, 12
gauges, 30
grease icon, 31

H

hand trucks, painting, 11
hardware categories, 18
heavy traffic, 6
historical data, predictive maintenance, 30
home locations, 5, 8

I

icons, 31
inaccessible places, 23
industrial-strength paint, 6
inventory
costs calculation, on-hand, 17
defined, 45
Kanban sizing report, 14
new quantities identification, 16
on-hand, 15, 17
overview, *xi, xii–xiii*
poor control, 21

J

jigs and fixtures, 2, 5
junk and garbage, 2
just-in-case scenarios, *xiii*

K

Kanban activities
benefits, 13–14
cards, 13, 19, 20
productivity, *xii*
quality, *xii*
sizing report, 14, 15, 17

L

labels, 11
Lean metrics
floor space, *xiii*
inventory, *xii–xiii*
overview, *xii*
productivity, *xii*
quality, *xii*
throughput time, *xiv*
travel distance, *xiii–xiv*
lock-out tags, 34
loosening/tightening, 30, 31
low-use items, 2
lumber mill examples, 38, 41

M

machine operator procedures and roles, 31–33
machine shop, 7–12
machine status, 39, 40
maintenance inventory
    categorizing, 18
    cycle counting, 15
    Kanban benefits, 13–14
    Kanban cards, 13, 20
    Kanban sizing report, 14, 17
    maximum quantity examples, 19
    new inventory quantities identification, 16
    new on-hand costs calculation, 17
    on-hand inventory, 15
    overview, 13, 21
    part identification, 15
    vendors, 15
maintenance parts room, 4
maintenance records review, 37
material categories, 18
Maximum Quantity, 16, 19
Minimum Quantity, 16
minor flaws, 23
motion, *xi*, 45

O

one-turn method, 25–27, 45
on-hand inventory, 15, 17
Ortiz, Chris, 49
overprocessing, *xi*, 45
overproduction, *xi*, 45
overview
    baselining equipment, 23
    5S activities, 1–2
    maintenance inventory, 13, 21
    total productive maintenance procedures, 29–30

P

painted work areas and aisle ways, 6
parts
    abnormality, 23
    categories, 18
    home locations, 5
    identification, 15
    Kanban cards, 19
    Kanban sizing report, 14
    maintenance inventory, 15
    organization basis, 20
    preventive maintenance, 30
    Red Tagging, 2
    standardization activities, 12
pictures, 20
predictive maintenance, 30
preventive maintenance, 30
proactive maintenance, 30
procedures
    machine operators, 31
    total productive maintenance, 12
    visual TPM boards, 39–40
Production Saw Change SWP, 38
productivity
    defined, 45
    5S activities, *xii*
    overview, *xii*
    total productive maintenance procedures, 12

Q

quality
    defined, 46
    5S activities, *xii*
    overview, *xii*
quick disconnect system, 25, 26

R

racks, 10
rails, wiping, 33
records review, 37
Red Tagging, 2–3, 46
rented container, 4
Reorder Quantity, 16
right sizing, 46
roll paint, 6
rotation, predictive maintenance, 30

S

sample icons, 31
scrubbing activities, 10–11, 46
service shop sort activities, 4
set in order activities, 5, 46
shadow board, 9, 46
sheet cycle counting, *see* Cycle counting
shelves, 10–11
shift recognition, 39, 40
shift walkthrough, 12
shop supplies categories, 18
shutdown tasks, 30
signs, 12
sizing report, *see* Kanban activities

sliding channels for panels, 26
solution-based engineering procedures, 36–37
sort activities, 3–5, 46
spray paint, 6
standardization activities, 12, 46
state of organization, 1, *see also* 5S activities
stencils, 11
storage bins, 10
supplies
   home locations, 5
   Red Tagging, 2
   shadow board, 9
   shop categories, 18
surfaces, scrubbing activities, 10
sustain activities, 12, 46

T

tables, scrubbing activities, 11
tape measure icon, 31
tasks, preventive maintenance, 30
throughput time, *xiv,* 46
tightening/loosening, 30, 31
tools
   boards, 12
   home locations, 5, 8
   preventive maintenance, 30
   Red Tagging, 2
   scrubbing activities, 10
   shadow board, 9
   standardization activities, 12
total productive maintenance (TPM)
   defined, 46
   fluid levels, 35
   lock-out tags, 34
   lumber mill example, 38
   machine operator procedures and roles, 31–33
   maintenance records review, 37
   overview, *vii–viii,* 29–30
   predictive maintenance, 30
   preventive maintenance, 30
   proactive maintenance, 30
   productivity, 12
   sample icons, 31
   solution-based engineering procedures, 36–37

total productive maintenance (TPM) boards
   attributes, 40
   defined, 46
   examples, 40–42
   materials to create, 40
   overview, 39–42, 46
tracking sheet, 5S activities, 45
transportation, *xi,* 46
travel distance, *xiii–xiv,* 46
turning operations, *see* One-turn method

U

unnecessary items, 23
unsafe places/conditions, 23

V

vacuum icon, 31
vendors, 14–15
visual TPM boards
   attributes, 40
   defined, 46
   examples, 40–42
   materials to create, 40
   overview, 39
visual workplace, 45
vital component abnormality, 23

W

waiting
   defined, 46
   overview, *xi*
   travel distance, *xiv*
walkthroughs, daily, 12, 45
wasted human potential, *xi,* 46
wastes, *x–xi*
window cleaning icon, 31
wiping rails, 33
work areas and aisle ways
   5S activities, 6
   set in order activities, 5
   waste of incorrect design, *xiv*
workbenches, 2

About the Author

Chris Ortiz is the founder of Kaizen Assembly, a Lean manufacturing training and implementation firm in Bellingham, Washington. Chris has been featured on *CNN Headline News* on *Inside Business* with Fred Thompson. He is the author of six books on Lean manufacturing (see list that follows).

Chris Ortiz is a frequent presenter and keynote speaker at conferences around North America. He has also been interviewed on KGMI radio and the *American Innovator* and has written numerous articles on Lean manufacturing and business improvement for various regional and national publications.

Kaizen Assembly's clients include industry leaders in aerospace, composites, processing, automotive, rope manufacturing, restoration equipment, food-processing, and fish-processing industries.

Chris Ortiz is considered an expert in the field in Lean manufacturing implementation and has over 15 years of experience in his field of expertise.

He is also the author of the following:

Kaizen Assembly: Designing, Constructing, and Managing a Lean Assembly Line (Taylor and Francis, 2006), now in second printing

Lesson from a Lean Consultant: Avoiding Lean Implementation Failure on the Shop Floor (Prentice Hall, 2008)

Kaizen and Kaizen Event Implementation (Prentice Hall, 2009); translated into Portuguese

Lean Auto Body (Kaizen Assembly, 2009)

Visual Controls: Applying Visual Management to the Factory (Taylor and Francis/Productivity Press, December 15, 2010)

The Psychology of Lean Improvements: Why Organizations Must Overcome Resistance and Change Culture (CRC Press and Productivity Press, April 2012): winner of the Shingo Prize for Operational Excellence in Research, 2013

The Lean Playbook Series (Taylor and Francis/Productivity Press, 2015–2016)